Microbes at Biomedical Interfaces 2018

Topical Conference at the 2018 AIChE Annual Meeting

Pittsburgh, Pennsylvania, USA
28 October - 2 November 2018

ISBN: 978-1-5108-7639-2

Printed from e-media with permission by:

Curran Associates, Inc.
57 Morehouse Lane
Red Hook, NY 12571

Printed by Curran Associates, Inc. (2019)

For permission requests, please contact AIChE
at the address below.

AIChE
120 Wall Street, FL 23
New York, NY 10005-4020

Phone: (800) 242-4363
Fax: (203) 775-5177

www.aiche.org

Additional copies of this publication are available from:

Curran Associates, Inc.
57 Morehouse Lane
Red Hook, NY 12571 USA
Phone: 845-758-0400
Fax: 845-758-2633
Email: curran@proceedings.com
Web: www.proceedings.com

TABLE OF CONTENTS

Some abstracts in this book refer to images that were not made available by the publisher

541771 Bad to the Bone: Biofilm and Surgical Infection

Abstract:

Infection remains the oldest and largest challenge in surgery. Knee and hip arthroplasty or total joint replacement provides an excellent illustration of this problem. Arthroplasty is one of the largest major surgical procedures in the world by volume and is an engineering marvel. It has been named one the greatest medical innovations of the twentieth century. Periprosthetic joint infection is the most common cause for total knee arthroplasty failure, and is a devastating surgical complication. Similar to other surgical site infections, morbidity and mortality is high. *Staphylococcus aureus*is the most common organism associated with surgical infections, including periprosthetic joint infection. These organisms can rapidly form a biofilm in the wound and on the implant. First-line treatment for surgical infection is irrigation, debridement, and treatment with long term antibiotics. Failure rate of irrigation and debridement is high. This high failure rate is a result of the high tolerance of biofilm to antibiotics. This tolerance is regulated by a number of different mechanisms. One of these includes bacterial persisters, a small subset of a bacteria population which are phenotypically tolerant to antibiotics. Persisters are clinically important because they are increased in biofilms, and play a role in the relapse of chronic infections. This is achieved, in part, through multiple toxin-antitoxin systems that induce a decreased metabolic state. A limited number of experimental methods to kill persisters have been discovered, and persister metabolism plays a vital role in each approach.

541824 The Microbiome in Health and Disease

Abstract:

Millions of microbes live in the gut and other body sites, yet are not able to be cultured and detected. Molecular techniques have recently been developed that can detect microbes at very low levels that exist in healthy individuals and in those with various diseases. These organisms, termed the microbiome, may have a significant impact on the development and progression of many diseases, and understanding these microbes may lead to new types of therapies for diseases such as cardiovascular disease, obesity, cancer, and autism.

The Human Microbiome Project was launched in 2008 as an effort to better identify microorganisms in the body of healthy individuals. Many other research projects are focused on determining the bacteria, viruses, and fungi in body sites such as the gastrointestinal tract, the skin, or the mouth in order to determine the relationship of these communities with development of different diseases. Up until a few years ago, it was thought that the lung didnâ€™t contain any microorganisms. However, it has been discovered that it does have its own microbiome in both health and disease. Significant changes in the amount or type of microbes in the lungs may cause inflammation and potentially lead to lung damage. Understanding noticeable changes in microbes may lead to new types of therapies to treat lung disease and related illnesses.

541651 Viable but Non-Culturable and Persistence Describe the Same Bacterial Stress State (invited talk)

Abstract:

Bacteria are often thought of as having two dormant phenotypes: the viable but non-culturable (VBNC) state and the persister state. Persister cells are a multi-drug tolerant subpopulation of bacteria that contribute to chronic and recalcitrant clinical infections such as cystic fibrosis and tuberculosis, and we have demonstrated that persister cells are metabolically dormant (Antimicrob. Agents Chemother. 57:1468–1473, 2013). VBNC cells are formed during stresses like extended starvation and are reported to not be resuscitated under normal conditions. We investigated the relatedness of the two stress-induced phenotypes at the single-cell level to examine cell morphology and quantify cell resuscitation (Environ. Microbiol. on-line 2018, doi: 10.1111/1462-2920.14075). Using the classic starvation conditions to create VBNC cells, we found that the majority of the remaining Escherichia coli K-12 population are spherical, have empty cytosol, and fail to resuscitate; however, some of the spherical cells resuscitate immediately. Critically, all the culturable cells in this starved population became persister cells within 14 days of starvation (similar results were found with the environmental isolate E. coli O157:H7 EDL933). We found that the persister cells initially are rod-like, have clear but limited membrane damage, can resuscitate immediately, and gradually become spherical by aging. After 24 h, only rod-shaped persister cells survive, and all the spherical cells lyse. Both cell populations formed under the VBNC-inducing conditions and the persister conditions are metabolically inactive. Therefore, the bacterial population consists of dead cells and persister cells in the VBNC-inducing conditions; i.e., the non-lysed particles that do not resuscitate are dead, and the dormant cells that resuscitate are persister cells. Hence, "VBNC" and "persister" describe the same dormant phenotype. In this presentation, we will also demonstrate that persister cells wake immediately when presented with nutrients and that they wake as a function of their ribosome content (Environ. Microbiol. on-line 2018, doi: 10.1111/1462-2920.14093).

541895 Mechano-Morphogenesis and the Capillary Peeling of Biofilms

Abstract:

Biofilms are surface-attached communities of bacteria that can damage industrial systems and cause serious infections in human health and fouling, while biofilms can be useful in wastewater treatment and microbial fuel cells. A comprehensive understanding of both the biological and physical properties of biofilms, and the way these properties interrelate, is necessary, though these properties are underappreciated. Here we use the model biofilm forming bacterium *Vibrio cholerae*, the causative agent of the pandemic disease cholera, to investigate the structure-property relationship of biofilm materials. Furthermore, by systematically quantifying the interfacial energy of *V. cholerae* biofilms, we identify capillary peeling as an efficient, easy technique to remove biofilms and to transfer biofilms from one surface to another without destroying their internal structures. Finally, we use the bulk rheological properties and the surface adhesion energy to rationalize the wrinkled shapes associated with various biofilm phenotypes, which here are given a mechanical interpretation.

540861 Biofilm Growth Drives the Selective Targets and Trajectories during the Evolution of Antimicrobial Resistance

Abstract:

*Acinetobacterbaumannii*is a persistent nosocomial pathogen categorized as one of the highest threats to patient safety. Part of its persistence is due to the ability to live in protected surface communities (biofilms) and evolve resistance to antibiotics, making the understanding of these evolutionary strategies a top public health priority. Using the susceptible *A. baumannii*ATCC 17978 as a model strain, we combined experimental evolution, antibiotic susceptibility tests, and whole genome sequencing to investigate different antibiotic-specific evolutionary pathways. We exposed both a biofilm and a well-mixed planktonic population to two different environments: i) sub-inhibitory concentrations of ciprofloxacin and ii) steadily increasing concentrations of ciprofloxacin. Our work demonstrates that bacteria evolve along distinct evolutionary pathways depending upon the mode of growth. Biofilm adapted lineages acquire mutations in three RND efflux pump regulators to mitigate antibiotic exposure whereas planktonic adapted lineages also acquire mutations in the target of ciprofloxacin, DNA gyrase. The biofilm-adapted lineages are more diverse, indicating spatial structure reduces clonal interference and can lead to greater odds of surviving fluctuating environmental stressors. Additionally, there is a trade-off between fitness cost and resistance level conferred by the adaptations in the different lifestyles and those lifestyle-dependent resistance pathways are related with cross-resistance and collateral sensitivity interactions between different antibiotics. We argue that the mode of growth (biofilm or planktonic) is an underappreciated yet key factor in the evolution of antibiotic resistance.

534490 Invited Talk 2: Repeatability of Metabolic Profiles in Multispecies Biofilms â€“ Toward Metrics for Biofilm Comparability

Abstract:

Novel materials and approaches to control microbial biofilms are typically tested *in vitro* to evaluate their efficacy. Consistency among control biofilms within an experiment, across experiments, and among laboratories and operators is needed to enable repeatability within a lab and reproducibility and comparability across labs. *In vitro* model biofilm systems are increasingly multispecies rather than monospecies to better represent naturally occurring biofilms. However, reproducibility in multispecies biofilms can be daunting due to their complexity including genotypic, phenotypic, physical, chemical, biological, and mechanical heterogeneities on multiple length scales. We hypothesize that biofilm function can be an indicator of biofilm repeatability and reproducibility, and metrics based on metabolomic studies can help establish criteria to demonstrate a biofilm is within expected limits, thus enabling comparison among laboratories.

The objective of this work was to evaluate the repeatability of metabolic profiles for a two-species model biofilm through small molecule mass spectral analysis. The biofilms consisted of *Pseudomonas aeruginosa* and *Staphylococcus aureus*, two common pathogens often co-located in human disease states including medical device infections, chronic wound infections, and chronic lung infections of cystic fibrosis (CF) patients. The species were inoculated as monocultures or â‰ˆ50:50 co-cultures and grown for 18 h in artificial sputum medium to mimic nutrient conditions found in CF lung infections. Experiments were performed on three separate days with three biological replicates and three technical replicates. Cells were quantified via plating on selective media and DNA extraction followed by quantitative polymerase chain reaction. Spent culture medium and cells were quenched and extracted to obtain exo-metabolites and endo-metabolites, respectively. Global metabolic profiles were obtained using liquid chromatography tandem-mass spectrometry (LC-MS/MS) and analyzed using in-house software platforms.

Co-cultures consisted predominantly of *P. aeruginosa*, as expected based on literature. All metabolic profiles had an acceptably low variability over three levels (day, biological replicate, and technical replicate). Principal component analysis indicated that metabolic profiles of *P. aeruginosa, S. aureus*, and the co-culture were statistically differentiable (n = 9), with exo-metabolic profiles of the co-culture having over 500 unique features as compared to monocultures. All cultures had over 500 compounds in common but at different quantities, suggesting up- and down-regulation of pathways in the co-culture. These results suggest that LC-MS/MS metabolic analysis holds promise to monitor repeatability and reproducibility in polymicrobial biofilms. Additional studies are needed to determine robustness including across laboratories, culture conditions, and timepoints, and also to identify unique features of interest. Overall, our approach has the potential to enable comparison and combination of data from multiple laboratories to improve characterization of material effects on biofilms.

540921 Invited Talk 3: Creating New Separation Processes By Interfacing Engineered Cells with Non-Living Material Interfaces

Abstract:

The creation of new biological interfaces has been a critical step for scientific exploration in the field of synthetic biology, yet these tools can also be used to create new chemical engineering technology with industrial relevance. One key area of interest is the separation and detection of biological and chemical species of interest. The detection of these species is complicated by both their low abundance in bulk fluids of interest, as well as a lack of sensitive instrumentation to detect these species. Thus, extraction, separation and/or pre-concentration are critical steps for detection. Here, we discuss how the tools of biointerface engineering and microfluidics can be used in sensing and separation. One long-term goal is to integrate the separation/concentration step and the signal amplification step in biosensing into one automated system, and we will discuss technologies we have developed to that end. We have developed initial standalone technologies for each component of such a system. We have created simple, low-cost, and robust mesofluidic handling systems for feeding fluid to microfluidic lab-on-a-chip devices (1). Additionally, we have created programmable material surfaces for detecting analytes synthesized by cells (2, 3). We will discuss preliminary data that demonstrates that these programmable materials can be miniaturized to create programmable microparticles that are sensitive to specific chemical and biological analytes. By capturing these particles in an in vitro, cell-free expression system, we have created the key enabling technologies to perform separation, capture, amplification and detection of chemical and biological species in a single system that will be low-cost, lightweight, low-power, and will have a long shelf-life â€" all characteristics ideal for new separation processes.Our long-term vision is that by using mesofluidic liquid handling and sampling systems to repeatedly introduce samples to a microfluidic separator, we will be able to handle bulk fluid samples, while also addressing low analyte concentration by binding key analytes with microscale sensors that do not require large amounts of analyte. The tools that we will present are the first steps to enabling technology towards this mission. They illustrate an industrially relevant payoff followingly the initial, hypothesis-driven development of biological interfaces between engineered living cells and non-living material systems.

References

1. J. R. Lake, Heyde, K.C., and Ruder, W.C., Low-Cost Feedback-Controlled Syringe Pressure Pumps for Microfluidics Applications. *PLoS ONE*, (2017).
2. K. C. Heyde, F. Y. Scott, S.-H. Paek, R. Zhang, W. C. Ruder, Using Synthetic Biology to Engineer Living Cells That Interface with Programmable Materials. e55300 (2017).
3. R. Zhang, K. C. Heyde, F. Y. Scott, S.-H. Paek, W. C. Ruder, Programming Surface Chemistry with Engineered Cells. *ACS Synthetic Biology*,(2016).

538323 Antifungal Peptide Variants with Reduced Degradation By Fungal Proteases and Improved Antifungal Activity Against Planktonic and Biofilm Cells

Abstract:

Histatin 5 is a highly cationic peptide found in human saliva that has specific and potent antifungal activity against the fungal pathogen *Candida albicans*. Although histatin 5 has potential as a therapeutic to combat *C. albicans* infections, *C. albicans* cells secrete proteases called secreted aspartic proteases (Saps) that can degrade and inactivate histatin 5. Two of the most important Saps are Sap2 and Sap9, due to their high levels of expression, and these enzymes cleave histatin 5 at its lysine residues. We designed analogs of histatin 5 with substitutions at its four lysine residues to study the effect on resistance to proteolysis by Sap2 and Sap9 and on antifungal activity. The single amino-acid substitutions of K17R and K17L were the most resistant to proteolysis, with 100% and 77% intact peptide remaining after incubation with purified Sap2 and Sap9, respectively, under conditions that led to only 61% (Sap2) and 47% (Sap9) of Hst-5 remaining. The decrease in proteolysis resulted mainly from reduced cleavage of the peptides on the C-terminal side of the K17 residue, and the decrease in proteolysis was also observed following incubation with *C. albicans* cells and culture supernatant. The protease-resistant variants of histatin 5 maintained the antifungal activity of the parent peptide, and the K17R, K17L, and K11R peptides showed significant increases in residual antifungal activity compared to histatin 5 following incubation with Sap2 and Sap9. By combining the K11R and K17R mutations, we were able to further improve the protease resistance and the antifungal activity of the peptides. The K11R-K17R variant had stronger antifungal activity than the parent peptide and maintained full antifungal activity following incubation with the proteases. The modified peptides also have improved activity against *C. albicans* biofilms and lack toxicity towards mammalian cells, further illustrating their potential application in designing improved therapeutics. We are currently exploring the incorporation of the histatin 5 variants into surface coatings to determine their potential in preventing biofilm-associated infections. Our results provide a unique perspective on antimicrobial peptide design by focusing on proteolytic resistance, in addition to antifungal activity, which could be particularly relevant in applications involving sustained contact with pathogen and host proteases.

540027 Effect of Poly-L-Lysine Molecular Weight on Antibacterial Activity of Polyelectrolyte Multilayer Coated Surfaces

Abstract:

Over 60% of hospital-acquired infections are device related. There is a need for device coatings that can inhibit bacteria attachment, preventing biofilms, while also killing planktonic bacteria. Small molecule antimicrobials embedded in such coatings can have uncontrolled release increasing the likelihood of drug resistance, increased localized toxicity, lack of stability, and loss of activity in the case of covalent attachment. Purely polymeric coatings can overcome many of these complications. We have developed such coatings using layer-by-layer self-assembly of hyaluronic acid (HA) and poly-L-lysine (PLL), a polycation, which plays both a structural and functional role due to its intrinsic antibacterial properties. An increased understanding of polymer physiochemical properties can improve their use in functional materials. We investigated the effect of PLL molecular weight (MW) on the antibacterial efficacy of (PLL/HA) films, finding that MW influences release-based killing while film hydrophilicity for all PLL MWs examined imparts resistance to bacterial adhesion. (PLL/HA) films were assembled containing 50 bilayers and denoted PLL^{n}, where n represents the number of repeat units in the PLL utilized in a given architecture. Films containing PLL with 30, 100, or 400 repeat units (6.3 kDa, 21 kDa, and 84 kDa, respectively) were investigated. These films were exposed to Staphylococcus aureus in the exponential growth phase. LIVE/DEAD staining and incubation on agar showed a significant reduction in bacteria attachment on all coatings compared with uncoated substrates. We hypothesize that this effect is related to the hydrophilicity of the films as evidenced by the large increase in film thickness upon dry film hydration in phosphate buffered saline (775%, 670%, and 266% for PLL^{400}, PLL^{100}, and PLL^{30}, respectively). Upon repeated exposure to fresh bacteria inoculum, PLL^{400} films lost bacterial growth inhibition activity rapidly (after 1 day), while PLL^{30} and PLL^{100} were effective for 4 days and approximately 10 days, respectively. We found that PLL concentrations were at or above the minimum inhibitory concentration of PLL for each MW against S. aureus only while films were effective, suggesting that growth inhibition occurs due to released PLL. Higher amounts of lower MW PLL were released into the media daily compared to larger MW PLL, likely related to a higher PLL mobility in films at lower MWs. Confocal imaging of films incubated with fluorescein labeled PLL confirmed that PLL mobility in the films decreased with increasing MW. PLL MW also influenced the thickness and stability of the films. PLL^{30} films (1.2 Â± 0.2 Î¼m thick) underwent complete dissolution by day 5, while PLL^{100} coatings (3.6 Â± 0.1 Î¼m thick) underwent disruption more slowly with considerable loss of film starting near day 10. This difference in stability contributed to the longer-term efficacy of PLL^{100} compared to PLL^{30}. Thus, we have demonstrated that PLL MW plays a critical role in dictating coating bacterial inhibition and long-term efficacy. The PLL films developed in this work may be tuned appropriately by varying PLL MW to prevent and treat device-associated infections.

533060 The Impact of Surface Topography on Adhesion and Biofilm Formation of Cyanobacteria

Abstract:

Biomass based renewable energy production is increasingly gaining attention due to the finite nature of fossil fuels and rapid increase in CO_2 levels in the atmosphere. In this regard, cyanobacteria (blue-green algae) based third generation biofuels are considered a promising alternative to fossil fuels, mainly because of their ability to capture CO_2 as a sole carbon source to produce fuels by photosynthesis. To date, a number of fuels and high value chemicals have been produced from cyanobacteria, but their cultivation in photobioreactors (PBRs) is challenging due to fouling on the transparent inner walls of the PBRs. Fouling can affect the light transmission efficiency of the PBR materials which in turn reduce the light availability to cyanobacteria cells and overall biomass generation and biofuel productivity. In order to prevent or reduce cyanobacteria fouling, we subjected polycarbonate PBR materials to surface topography modification by direct laser interference patterning (DLIP), and investigated the antifouling ability of the modified materials using a selected cyanobacterium (genetically engineered to enhance biofuel production). Various experimental techniques were used to characterize the biofilm formation of cyanobacteria on topography modified materials, including scanning electron microscopy (SEM), contact angle measurements, bright field microscopy, attenuated total reflectance-Fourier transform infrared (ATR-FTIR) spectroscopy, UV-Vis spectroscopy and atomic force microscopy (AFM). The feature size and topography of the modified PBR materials were examined by AFM. The transmission efficiency of the PBR materials was reduced by 5 - 11% as a result of laser micro patterning. The biofilm formation assay revealed that, after 7 days, the mean percent area coverage of cyanobacteria was reduced by 40% on topography modified PBR materials when compared to the smooth control surface. The cyanobacteria cells showed a tendency to align across the features, when the feature size of the topography was greater than the size of the microorganisms. Surface hydrophobicity of all the patterned materials was reduced within 4 days due to the adhesion of cyanobacteria as determined from the contact angle measurements. ATR-FTIR spectroscopy revealed proteins and polysaccharides as prominent groups involved in fouling of cyanobacteria on patterned PBR materials. Overall, the PBR materials modified through DLIP in this study showed an important role of surface topography in the prevention of cyanobacteria biofilms, indicating a viable approach to engineering antifouling materials by using this technique to explore a wide range of different topographical features and sizes.

Keywords: Biofouling, Photobioreactors, Cyanobacteria, Direct Laser Interference Patterning, Antifouling.

524862 Sensitizing Bacterial Cells to Antibiotics through Dynamic Topographyâ€“Triggered Biofilm Detachment

Abstract:

Microbial biofilm is a leading cause of chronic infections in humans and persistent biocorrosion and biofouling in industries due to extremely high-level tolerance of biofilm cells to antimicrobial agents. In the past three decades, the significance of biofilms has stimulated extensive studies on strategies for biofilm prevention and removal. However, due to the protection of extracellular matrix and slow growth of biofilm cells, controlling mature biofilm remains challenging. Recently, we reported that mature biofilms can be effectively removed (e.g. 99.9% detachment of 48h *Pseudomonas aeruginosa* PAO1 biofilms) using shape memory polymer-based dynamic change in surface topography. Here, we demonstrate that such biofilm detachment also sensitizes biofilm cells to conventional antibiotics. For example, concurrent treatment of 48h *P. aeruginosa* PAO1 biofilms during dynamic biofilm removal increased the susceptibility of released biofilm cells to tobramycin by more than 3 logs (2,479 fold) compared to surface-attached biofilm cells. The observed effects were attributed to the disruption of biofilm structure and increase in cellular activity as evidenced by a 11.8-fold increase in ATP level and 4.1-fold increase in expression of *rrnB* gene in detached cells. These results can guide the design of future combinational therapies for biofilm removal by integrating newly-designed smart dispersion strategies with currently available antibiotics.

536323 The Role of Flagellar Motor Reversals in Swarming in *Escherichia coli*

Abstract:

Swarming in thin fluid films on top of solid surfaces is enabled by the rotation of flagellar filaments in bacteria. To swarm, cells must overcome an increase in viscous drag due to the proximity to the surface. Previous reports suggest that cells likely compensate for the surface drag by promoting surface wetness through an unknown mechanism with the aid of a flagellar switch. The switch helps the filament erratically reverse the direction of rotation between clockwise (CW) and counterclockwise (CCW). The direction of rotation is dependent on the conformation of a switch-component, FliG. Here, we investigated the role of FliG conformation in facilitating bacterial swarming. We observed that swarming was abolished when FliG was locked in the CW conformation ($FliG^{CW}$), but not in the CCW conformation ($FliG^{CCW}$). Swimming was not abolished, so the conformation of FliG is likely critical for swarming. We found that torque generation in $FliG^{CW}$ strains was similar to WT strains. Swimming speed analysis revealed the loss of swarming ability correlated with a reduction in the flagellar thrust in the CW strain, which was likely due to a reduction in the number of filaments per cell. Fluorescence visualization showed that $FliG^{CW}$ swimmers have one less filament per cell on average compared to WT swimmers. Yet, $FliG^{CW}$ swarmers tend to lack filaments entirely (~57% of cells visualized carried no filaments, compared to ~10% in the WT). We hypothesized that there was a link between the function of the switch and degradation of the flagellar assembly; however, TIRF imaging revealed no significant degradation. Although the role of the switch in swarming remains unclear, our assays revealed that WT swarmer cells show a significant decrease in the probability of CW rotation (known as CW_{bias}) suggesting modifications are occurring at the level of the switch, and further pointing to its importance.

528605 Pseudomonas Aeruginosa Single-Cell Level Heterogeneity, Investigated Via Drop-Based Microfluidics

Abstract:

Single cell microbial studies can be challenging using traditional microbiology methods; however, single cell studies are critical in understanding cell-to-cell heterogeneity for interpreting bulk microbial growth behavior. This is especially true in biofilms, or microbial communities that express high levels of population heterogeneity and exceptional resilience due to its heterogeneity. Drop-based microfluidics, or the formation and manipulation of picoliter-volume drops of water-in-oil allows for the encapsulation of single cells, providing a platform for single cell studies. In this work, populations of Pseudomonas aeruginosa are examined at the single cell level within microfluidic drops incubated on a microfluidic chip over a period of 24+ hours. Confocal microscopy tracks the expressed plasmid gfp fluorescence of the cells. The resulting data is used to construct growth curves starting from single cells and to quantify the growth rate and lag phase. Our long-term growth study is performed using a device preparation technique that fills the porous microfluidic device with water-saturated carrier oil and allows for thermodynamic stability between the oil and aqueous phases. In absence of this preparation, drops are observed to shrink, foam, or evaporate from the device over time. Bulk studies of P. aeruginosa cells reintroduced to media after 4 days of starvation shows decreased growth in resuscitating cells. However, single cell studies using this incubation method show that these cells have increased and variable lag phase lengths, rather than decreased specific growth rates (Akiyama, et.al. PNAS, 2017). Our results demonstrate the quantification and monitoring of single cell behavior over 24+ hours, which will allow for a deeper understanding of the role of heterogeneity in bacterial populations.

537121 Differential Response of Mucoid and Non-Mucoid *Pseudomonas AeruginosaÂ* isolates to Interfacial Confinements

Abstract:

Fluid interfaces are energy-rich environments known to influence the self-assembly of small molecules or microparticles, and the formation of thin films by biological entities. The formation of bacterial films at fluid interfaces appears to be a dynamics process, in which cells coping with the existing interfacial energies respond to the entrapment by forming a matrix appropriate to their survival. We hypothesize that cells subjected to interfacial stress adapt and exhibit phenotypic changes essential to their survival. In this study, we are investigating the response to interfacial confinement of two clinical strains of *Pseudomonas aeruginosa* isolated from the airways of cystic fibrosis patients. To evaluate what phenotypic changes at fluid interfaces provide growth advantage, we are comparing a mucoid (PASL) and non-mucoid strain (PANT) of *P. aeruginosa* secreting various level of polysaccharides. We are investigating the mechanical properties of films formed by these cells to understand how they adapt and restructure environments with high interfacial energies. Some intrinsic properties of the cells confer viscoelastic behavior facilitating their adaption as characterized by the apparent moduli of the elastic films observed using pendant drop elastometry and particle tracking. Further characterization of whole genome and the transcriptomes of cells under interfacial confinements will provide new insights on the biological implications of interfacial films.

540115 Bloodmeal-Induced Inhibition of *Plasmodium* infection in Mosquito Vectors Using the Microbial Symbiont *Asaia*

Abstract:

Asaia sp. are gram-negative bacteria that have been shown to colonize the *Anopheles* mosquito midgut. Species of this mosquito are the vectors for malaria transmission throughout the world. Using a paratransgenic strategy in which transgenically modified symbiotic organisms affect their hostâ€™s phenotype, *Asaia* has been engineered to secrete anti-malarial effector molecules into the mosquito midgut, hindering its ability to carry *Plasmodium*. However, the constitutive expression of these effectors causes a fitness disadvantage to the transgenic *Asaia*.Therefore, it is desirable to express these molecules only when *Plasmodium*is present in the mosquito midgut, namely when a mosquito takes an infected blood-meal. Putative conditional promoters have been cloned into the plasmid pGLR1, which has a promoterless GFP reporter. These plasmids were then transformed into the lab strain *Asaia sp.SF2.1* and plated on minimal media enriched with blood. GFP fluorescent colonies were collected and screened both in-vitro, in liquid and solid media, as well as in-vivo, inside the mosquito midgut, to isolate the promoters that were only induced when blood meal conditions were present. Once the conditionality of these promoters was validated, they were cloned into a plasmid secreting the anti-malarial effector scorpine and transformed into *Asaia*. The increased fitness of these conditional strains was validated against a constitutively expressing control before testing each strain for *Plasmodium* inhibition. The anti-malarial strains were cultured in mosquitoes that were then fed on *Plasmodium*infected mice. The conditionally active promoter strains showed a significant reduction from the wild-type bacteria in the number of oocysts the pathogen formed as well as a significant reduction in disease prevalence effectively inhibiting the mosquitoesâ€™ ability to transmit malaria.

537630 Dynamics of Biofilm Elimination on Thermally Shocked Biomedical Surfaces

Abstract:

When bacteria colonize a medical implant surface, they form a biofilm which cannot be eradicated chemically. The current standard of care is surgical explantation of the device and surrounding tissue, with eventual reimplantation of a replacement device with twice the probability of infection. These infections are a $5 billion problem in the U.S. alone, impacting over 100,000 patients annually. One approach to mitigating these biofilms is *in situ* thermal eradication by generating a thermal shock directly at the device/biofilm interface. We have developed magnetic nanoparticle / polymer composite coatings which heat rapidly when exposed to an externally applied alternating magnetic field, and have quantified immediate bacterial population reductions of up to 6 orders of magnitude (i.e., "6-log" reductions).

Moreover, under certain circumstances the biofilm bacterial population continues to decrease for hours after the thermal shock is removed and physiological temperature is restored, resulting in complete eradication of the biofilm. Re-incubation studies at a variety of shock temperatures (50 – 80 °C) and exposure times (1 – 30 min) showed that thermal shocks producing immediate population reductions greater than 4-log resulted in complete elimination of the biofilm approximately 4 hours later, regardless of the temperature/exposure time combination used to achieve the immediate reduction. Biofilms subjected to thermal shocks with more modest immediate population reductions (less than 4-log) grew back to their initial population density within 24 hours. These investigations were extended to biofilms with different culture conditions producing pre-shock population densities 100 times greater than the earlier biofilms (nearly 10^9 CFU/cm^2). Despite the substantially larger initial population density, the same immediate population reduction (4-log) was required to prompt eventual die-off of the biofilm, even though the population density immediately after the thermal shock was 100 times greater than in the previous studies. This suggests that the degree of thermal shock required to eradicate a biofilm is independent of the biofilm's initial population density.

Furthermore, this approach may be integrated with other modifications to the biofilm/device interface to prevent or eliminate biofilms with even milder thermal shocks. Investigations of combined thermal shock and antibiotic exposure yielded population reductions whose orders of magnitude were greater than the sum of the orders of magnitude reduction for either treatment alone, enabling significantly milder thermal shocks to provide the same efficacy and providing an impetus for localized delivery at the device interface of antibiotics which are by themselves insufficient for biofilm elimination.

535582 Modelling Microbial Microenvironments through Encapsulation of Synthetic Communities

Abstract:

In the field of microbiology, it is believed that less than 1% of microbes are presently known and have been studied. The inability to culture many microbes using traditional methods, such as plating or suspension, has led to a vast gap in our understanding of microbiomes. To overcome this challenge, we have developed a novel microfluidics system which enables us to encapsulate microbes in a semipermeable membrane on the scale of nanolitres, eliminating resource competition and effectively increasing cell communication of isolated microcolonies.

In this study, we are introducing a mock community developed in the nanocultures with five microbial species, including *Pseudomonas aeruginosa, Escherichia coli Nissle 1917, Staphylococcus aureus, Streptococcus mutans and Candida albicans*. We determine how population density differs when cells are cultured in nanocultures as opposed to culturing them in suspension in a flask. After an incubation period of 24 hours, the nanocultures are ruptured and DNA is isolated and sequenced from viable cells. 16srRNA sequencing allows us to determine the abundance of each species in both the nanocultures and flask cultures, thus we can determine how the species interact with each other under the constraints of each system. We anticipate that our microfluidics device will be able to capture unculturable microbes in future applications, from a variety of different environments, such as oral, gut, marine and soil, allowing for unknown microbe discovery, and furthermore, a more accessible and in-depth study of these unknown microbes.

540575 Control of Pseudomonas Aeruginosa Biofilms By Electrical Currents Using a Simple Agar Model

Abstract:

Electroceutical wound dressings show promise for treating biofilm infections, however, the mechanism of action is poorly understood. We developed an in vitro agar based wound infection model using a bioluminescent strain of Pseudomonas aeruginosa to measure loss of activity and killing when direct current was applied to silver electrodes overlaid with agar. A 6V battery with 1kÎ© ballast resistor was used to expose the biofilms to the electric current for 1h and 24h. Loss of bioluminescence and a 4-log reduction in cells was achieved over the anode. Scanning electron microscopy showed that the biofilm had been disrupted and cells showed evidence of lysis. Even after turning off the current the antimicrobial activity continued to spread from the anode for 3 days suggesting generation of a stable antimicrobial agent had occurred. Elevated levels of chloride in the agar and chlorotyrosine in the dead biofilm suggested a role for hypochlorous acid. Similar killing was obtained for biofilms grown on gels containing onlybovine synovial fluid or human serum. We hypothesize that in the presence of physiological fluids, hypochlorous acid is produced at the anode which may react with proteins in the agar and/or the biofilm to form stable reactive intermediates.

524748 Prevention of Select Eskape Pathogens from Attaching to Titanium Using Cathodic Voltage Controlled Electrical Stimulation Combined with Antibiotic Therapy

Abstract:

INTRODUCTION: Infection following total joint arthroplasty (TJA) has been associated with longer patient healing times, prolonged hospital stays and increased costs to the healthcare system. ESKAPE pathogens are a concerning class of multidrug resistant bacteria in the medical setting that are contributing to these infections. In a previous study, we revealed that cathodic voltage-controlled electrical stimulation (CVCES) could prevent coupon-associated and planktonic colony forming units (CFU) of Gram-positive methicillin resistant *Staphylococcus aureus* and Gram-negative *Acinetobacter baumannii* from reaching detectable levels in a magnitude and time-dependent manner [1]. In this study we evaluated if application of CVCES at -1.0V, -1.5V, or -1.8V vs. Ag/AgCl, in conjunction with prophylactic antibiotics, is an effective method to prevent attachment of *Staphylococcus epidermidis* (*S. epidermidis*), *Klebsiella pneumoniae* (*K. pneumoniae*), *Acinetobacter baumannii* (*A. baumannii*), and *Pseudomonas aeruginosa* (*P. aeruginosa*) to the surface of commercially pure titanium (cpTi).

METHODS: cpTi coupons were wet sanded to a 600-grit finish, sonicated in deionized water for 10mins, and sterilized under ultraviolet light for 30mins. Experiments were conducted with blood isolates of Gram-positive *S. epidermidis*, strain SE94-309-0594, and Gram-negative *A. baumannii*, strain Ab307-029, *K. pneumoniae*, strain cKP1, and *P. aeruginosa*, ATCC(T) strain PA27853. Fresh cultures were inoculated into 5 mL of either tryptic soy broth supplemented with 0.25% glucose (TSBG, for SE94-309-0594) or Mueller-Hinton (MH) broth (Ab307-0294, cKP1, PA27853) at a 10:1 flask to volume ratio, adjusted to an OD_{600} of 0.1 and placed in a 37°C shaking water bath at 180 rpm. Cultures were grown to mid-log phase, diluted back to an OD_{600} of 0.1 and then further diluted 1:10,000 in the appropriate fresh media for a starting inoculum of approximately 10^3 CFU/mL. Minimum inhibitory concentrations (MICs) of antimicrobial agents were determined by the broth dilution method recommended by the Clinical and Laboratory Standards Institute (CLSI) [18]. The MIC for cefazolin against SE94-309-0594, Ab307-0294, and cKP1 was determined to be 1 µg/mL, 400 µg/mL, and 2 µg/mL, respectively. MIC of gentamicin was tested against PA27853 and found to be 3 µg/mL. Titanium was incubated for 24-hours at open circuit potential (OCP), OCP+drug, CVCES alone, or CVCES+drug. The stimulation chamber uses a standard three-electrode system with cpTi as the working electrode, a graphite counter electrode and an Ag/AgCl reference electrode [2]. With respect to Ag/AgCl, voltages of -1.0V, -1.5V, and/or -1.8V for 24 hours were applied to the working electrode at room temperature using a potentiostat (Interface 1000, Gamry Instruments). After incubation, cpTi coupons were washed in sterile PBS, sonicated and dilution plated for CFU. The media was collected for pH measurement and dilution plated for CFU. Four independent samples were evaluated in all conditions. A one-way ANOVA followed by Tukey post-hoc compared log-transformed CFU across stimulation conditions. If the Levene's test was significant, a Welch's test followed by a Games-Howell post-hoc test was performed. Student's t-tests were used to assess differences between independent stimulation conditions with and without antibiotic treatment.

RESULTS: Remarkably, CVCES of *P. aeruginosa* at -1.0V+gentamicin and -1.5V+gentamicin showed an 8-log reduction, to below detection, from controls and CVCES alone. -1.0V+cefazolin significantly reduced coupon-associated *A. baumannii* from control ($p<0.001$), and -1.5V resulted in no detectable coupon-associated CFU and reduced planktonic CFU to below 10^1 CFU/mL. Compared to controls -1.8V and -1.8V+cefazolin prevented all *K. pneumoniae* from reaching detectable levels ($p<0.001$), but had no effect at -1.5V and -1.5V+cefazolin. -1.0V+cefazolin to *S. epidermidis* reduced planktonic CFU by 93.2% and resulted in no detectable coupon-associated CFU. -1.5V alone significantly reduced *S, epidermidis* coupon-associated and planktonic CFU from controls ($p<0.005$), and prevented all CFU at -1.5V+cefazolin (data not shown). Increasingly cathodic CVCES had a higher average cathodic current density and cumulative charge transfer. CVCES can prevent coupon-associated and planktonic CFU of select ESKAPE pathogens from reaching detectable levels on cpTi, and antibiotics synergistically enhance these effects.

DISCUSSION: Implant associated infection following orthopedic intervention is a devastating clinical complication associated with increased patient morbidity, extended treatment times, and increased costs to the health care system. The purpose of this *in vitro* study was to evaluate if CVCES of titanium in conjunction with antibiotics could synergistically prevent initial attachment and/or reduce the viability and subsequent bacterial biofilm formation of select ESKAPE pathogens. The outcomes of this study revealed that CVCES could prevent coupon-associated CFU and planktonic CFU of select ESKAPE pathogens from reaching detectable levels in a magnitude-dependent manner. In general, increasingly cathodic potentials had more robust antimicrobial effects and this effect was enhanced in the presence of antibiotic treatment. Bactericidal activity of stimulation corresponded with an alkaline shift in pH, which is thought to be the causative link between stimulation and bacterial cell death. In general, the addition of clinically relevant antibiotics enhanced the efficacy of CVCES to have a more robust preventative effect and these effects were dependent on the magnitude, antibiotic, and species being tested.

SIGNIFICANCE: Current antibiotic prophylactic treatment methods for joint revision surgeries are not effective against all pathogens; however alternative methods, such as CVCES in conjunction with antibiotics, shows promise to be an effective broad-spectrum prevention method for Gram-positive and Gram-negative pathogens.

REFERENCES: [1] Canty M, *et al.* Acta Biomaterialia: 2017;48:451-460 [2] Ehrensberger MT *et al.* Biomaterials 2015;41:97-105.

537838 Computational Modeling of Cathodic Voltage Controlled Electrochemical Treatment of Biofilms in-Vivo

Abstract:

Computational Modeling of Cathodic Voltage Controlled Electrochemical Treatment of Biofilms In-vivo

Amir Mokhtare[1], Mark Ehrensberger[2]and Edward P Furlani[1,3]

[1]Dept. of Chemical and Biological Engineering,[2] Dept. of Biomedical Engineering, [2] Dept. of Biomedical Engineering, [3]

University at Buffalo, The State University of New York, Buffalo NY

Email: efurlani@buffalo.edu

Abstract

The last few decades have witnessed a proliferation in the use of prosthetic implants. Chronic bone and joint related disorders such as osteomyelitis, septic arthritis, etc. are widespread and can have a devastating impact on quality of life. Consequently, the use of total joint arthoplasty (TJA) has grown steadily as a life enhancing procedure. Although successful joint replacement restores functionality and provides pain relief, prosthetic joint infections (PJIs) can be a devastating outcome of orthopedic surgery and can impose horrendous physical burdens on individual patients and financial stress on the health industry as a whole. PJIs are often caused by aggregated communities of bacteria known as bacterial biofilms, which tend to be incredibly resistant to antibiotics. As a result, new methodologies are needed to efficiently treat biofilm-based PJIs in-vivo. In this regard, electrochemical treatment of the infected implants is among the most promising strategies for the eradication of surface-based biofilm infections. Cathodic voltage controlled electrical stimulation (CVCES), is a recently developed electrochemical approach that has proven to be effective in the treatment of biofilm-based PJIs.

Analysis of in-vivo assays have revealed that CVCES combined with conventional antibiotic drugs can lead to complete treatment of the infections without any histological damage to the surrounding tissue. However, many fundamental aspects of this treatment are unknown and rational design towards optimization is lacking. In this presentation we introduce a one-dimensional (1D) computational modeling framework to simulate the electrochemical behavior of an in-vivo CVCES system. We use dilute solution theory (ionic interaction are neglected) for determining the electrochemical potential of ionic species in the system and solve the Nernst-Plank equations along with Maxwell's equations to predict the toxic (to bacteria) species concentration speciation and electric potential. Water-based electroneutrality is used as a simplifying assumption to facilitate numerical convergence. In addition, Butler-Volmer electrochemical surface reaction kinetics are assumed at the surface of the working electrode and as boundary conditions for the models. We have used our experimental current-voltage scanning data to identify the dominant electrochemical surface reactions and have extracted transfer coefficients and exchange current densities required to characterize the reactions. The surrounding tissue, i.e. the electrolyte medium in the in-vivo experiments, was approximated by the dilute solution of sodium chloride. Moreover, the most important buffer systems inside the tissue, namely the bicarbonate buffer system and buffer capacity of proteins and organic phosphates, were introduced in the form of homogeneous reactions. Hydrogen evolution and oxygen reduction electrochemical reactions are assumed at the working electrode surfaces and characterized by voltage current scanning data. Both experimental and mathematical modeling results showed that oxygen reduction surface reaction become diffusion limited after few minutes and only contribute a small amount to the total current density at the electrode surface. Simulation results also elucidate how buffering capacity of the tissue would counteract the diffusion of hydroxide ions in the tissue. The mathematical model also provides the temporal-spatial variation of the both current-density and pH profiles that are very difficult to obtain in experiments. The simulation results are in an acceptable quantitative range with the experimental results. The modeling framework can be extended to higher dimensions to represent more realistic examples such as human knee or hip implants. Such models can be utilized to optimize the electrode configurations and operation parameters.

Keywords: Biofilm induced infections, prosthetic joint replacement, in-vivo mathematical modeling, electrochemical treatment. Results

Figure1. Schematic of the in-vivo assay and computational domain, a) pH profile distribution

b) concentration profiles c) IV experimental and modeled data d) transfer coefficients of hydrogen evolution reaction

References

[1] M. T. Ehrensberger *et al.*, *Biomaterials,* vol. 41, pp. 97, 02/ 2015.

[2] F. L. Hellweger *et al.*, *Nature Reviews Microbiology,* vol. 14, no. 7, pp. 461, 2016.

[3] E. Nilsson *et al.*, *Bioelectrochemistry,* vol. 53, no. 2, pp. 213, 2001.

[4] S. Nodzo *et al.*, *Clinical Orthopaedics and Related Research®,* vol. 473, no. 9, pp. 2856, 2015// 2015.

538388 Electroactive Surfaces and Their Use for Biofilm Removal to Advance Wound Healing

Abstract:

Biofilm infections are present in nearly all chronic wounds, contributing to delayed wound healing. Biofilms are difficult to treat due to their dense structure, slow growth rate, altered metabolism, and protection by extracellular polymeric substances. Current approaches to biofilm infection treatment use a combination of debridement, and systemic and topical antibiotics. These approaches partially treat biofilm infections. However, their performance is hindered due to the depletion of active compounds at deep layers of biofilms and the relative non-susceptibility of biofilm bacteria to conventional antibiotics. This gives rise to persister cells that survive treatment, and leads to infection recurrence. In this work, we present a novel electroactive material surface and an electronic platform for the management of wound biofilm infections by continuously producing effective concentrations of antimicrobials near the electroactive material surface. The concentration of antimicrobials is tuned using a miniaturized electronic device so that the concentration is sufficient to eliminate biofilm infections without apparent damage to host tissue. The efficacy of this strategy was evaluated for the treatment of common biofilm forming pathogens using *in vitro* and *ex vivo* biofilm models. We used mathematical modeling with finite-element analysis software COMSOL Multiphysics® to predict the antimicrobial generation rate near electroactive material surfaces.

539225 Toward to the Design of an Electrochemical Therapy (ECT) Against Microbial Infection

Abstract:

One of the main challenges healthcare faces is the preponderance of antibiotic-resistant microorganisms in public and medical environments. The antibiotic-resistant species constitute significant challenges to curing chronic infections, especially those associated with implanted medical devices. We explore the antimicrobial activity of low-level electrochemical treatments on a range of microorganisms. We report that these pathogenic microbes can be effectively eliminated by low-level direct currents (DCs); e.g. treatment with 70 Î¼A/cm^2 DC for 1 hr using titanium or stainless steel (SS) 304 electrodes reduced the number of viable planktonic *Streptococcus mutans*, *Candida albicans*, *Pseudomonas aeruginosa* by 1-2 logs, compared to the untreated control. In addition to killing by applying DC alone, synergistic effects are achieved when treating the planktonic and biofilm cells with 70 Î¼A/cm^2 DC and antibiotic dosages concurrently. Interestingly, alternative ECT utilizing DC generated with graphite followed sequentially by 1.5 Î¼g/mL Tob, or DC via TGON 805 (a carbon-based material) as a sole factor, eradicated the persisters completely (~7-log). Electrochemical factors including electrode composition, electrochemical byproducts, and current/potential levels contributed to impairment of persistersâ€™ electrophysiology, causing substantial changes in cell structures and transcriptomes. This work will define the conditions necessary to modulate electrochemically-induced sensitization of pathogens and eradicate polyspecies biofilms from implantable materials.

530184 Wireless Electrostimulation to Eradicate Bacteria Biofilm

Abstract:

Pathogenic bacteria such as *Pseudomonas aeruginosa* and *Staphylococcus aureus* can form biofilms on medical implants and cause serious infections that are incurable by conventional antibiotics due to high-level drug resistance. As an alternative approach, electric currents have been shown to kill biofilm cells and have synergy with antibiotics in bacteria killing. However, conventional electrochemical treatment requires connection electrodes to a power source. This requires invasive procedures and can cause complications, hindering the application of this technology *in vivo*. In this study, we demonstrate that biofilm cells can be effectively killed by using electromagnetic induction to deliver direct current wirelessly from a remote power source. For example, after treatment with 60 $\mu A/cm^2$ of wirelessly delivered DC conducted by stainless steel electrodes for 6 hours, the viability of biofilm cells was reduced by approximately 4 logs and 2.2 logs for *P. aeruginosa* and *S. aureus*, respectively. Similarly, the viability of *P. aeruginosa* and *S. aureus* biofilm cells was reduced by 2 and 2.5 logs after treatment with the 30 $\mu A/cm^2$ of wirelessly delivered DC conducted by graphite-based electrodes for 3 hours, respectively. Synergy in concurrent treatment with lower level DC and antibiotics to biofilms was also observed. These conditions appeared to be safe to the human epithelial cells. Encouraged by these findings, we engineered a model that contains the wireless DC treatment units and shares similar profile of implantable medical devices. The device was found to effectively kill of *P. aeruginosa* and *S. aureus* biofilms without noticeable toxicity to human cells. With the capability to kill bacteria without using a directly connected power source, this platform has potential applications in engineering smart devices to non-invasively eradicate biofilm infections.

527193 Electrochemical Detection of Bacterial Biofilms on Titanium

Abstract:

Electrochemical Detection of Bacterial Biofilms on Titanium

Caelen M. Clark and Mark T. Ehrensberger

State University of New York at Buffalo, Buffalo NY, USA

Statement of Purpose: Periprosthetic joint infection (PJI) is a devastating complication of total joint arthroplasty. In addition to being extremely difficult to treat with antibiotics alone, PJI is also difficult to detect [1]. These problems stem from the ability of bacteria to form biofilms on the surface of implanted materials. When in the biofilm state, bacteria gain a diffusional barrier which limits the penetration of antibiotics, and causes a reduction in the metabolic activity of the resident cells. This can make detecting biofilms with traditional culture methods difficult. For this reason, new methods for the detection of biofilms on implanted orthopedic materials are needed. This work evaluated the electrochemical methods of potentiometry and electrochemical impedance spectroscopy (EIS) as diagnostic measures of bacterial biofilm formation on titanium in an *in vitro* model.

Methods: A custom designed polycarbonate electrochemical biofilm reactor (Fig 1) was utilized to both grow biofilms and make the electrochemical measurements.

Figure 1: Experimental setup of the biofilm reactor

Samples were made from grade 4 commercially pure titanium (Ti). A clinical isolate of Acinetobacter baumannii (strain Ab307), a problematic Gram-negative human pathogen, was used in all experiments. Biofilms were grown by introducing a 30 mL inoculum with a concentration of $\sim 1\times 10^7$ CFU/mL to the reactor. After two hours, flow in the system was initiated at 0.02 mL/s. The reactor also functioned as a three electrode electrochemical cell utilizing the Ti as a working electrode, an Ag/AgCl reference electrode, and a graphite counter electrode. The open circuit potential (OCP) of the Ti was measured prior to inoculation of the reactor, and 24 and 48 hours after inoculation. EIS was performed using a Gamry Interface 1000 potentiostat with a 10 mV sinusoidal voltage about the OCP in the frequency range of 100 kHz-5 mHz at the same timepoints as the OCP measurements. The results of the EIS spectra were fit to a modified Randles circuit, and the polarization resistance (Rp), constant phase element (CPE) magnitude (Y0), and CPE exponent (α) were calculated. Following EIS, the Ti coupons were extracted, rinsed in phosphate buffered saline, sonicated in 0.1% saponin, and dilution plated for enumeration of CFU. Scanning electron microscopy (SEM) of the extracted Ti was also performed for qualitative assessment of bacteria on the Ti surface. Control experiments were carried out with the same experimental conditions without bacterial inoculation. A total of 9 samples used for both the biofilm and the control experiments, and a T-test was used to compare the electrochemical parameters at each time point.

Results: Biofilm formation on Ti resulted in a significant cathodic shift in the OCP after 24 hours as compared to the control samples. The Rp was found to increase significantly after 24 hours of biofilm growth (Table 1). CFU enumeration and SEM imaging (Fig 2) confirm the presence of biofilms with an average of 1.1 $\times 10^7$ CFU/mL.

	OCP (mV)		
Time	**Ab307**	**Control**	**P-Value**
0	-326±32		
24	-387±23	-321±20	<0.001
48	-381±21	-308±32	<0.001
	Rp (Ω/cm^2)		
	Ab307	**Control**	**P-Value**
0	$6.98\times10^6 \pm 4.6\times10^6$		
24	$5.4\times10^7 \pm 2.8\text{x}10^7$	$1.0\text{x}10^7 \pm 6.0\text{x}10^6$	<0.001
48	$1.1\times10^8 \pm 1.1\text{x}10^8$	$1.8\text{x}10^7 \pm 2.8\text{x}10^7$	0.04

Table 1: Electrochemical measurement results

Figure 2: SEM micrograph of Ab307 biofilm

Conclusions: Biofilms grown on Ti for 48 hours resulted in a statistically significant cathodic shift in the OCP of the metal at 24 and 48 hours compared to control samples. The Rp was also found to increase significantly compared to control samples after 24 hour of biofilm growth. These results show the effect of biofilm formation on the electrochemical properties of Ti. Further work will be conducted to track these electrochemical parameters as a function of biofilm formation and growth.

References: [1] Sendi P, Zimmerli W. International Journal of Artificial Organs. 2012;35:913-22.

534597 Novel Focused Multivector Ultraviolet (FMUV) Disinfection without Manual Cleaning and Chemical Disinfection in-between Surgeries and throughout the Hospital Environment

Abstract:

PurpleSun is an early stage healthcare technology company based out of New York City. We are focused on providing smart, light-based technology solutions to the healthcare industry for infection prevention, diagnostics and treatment.

Our companyâ€™s novel light technology platform will be utilized throughout the continuum of care, within hospitals, long term care and surgical centers. The smart devices will set the stage for patient safety through the highest of quality in medical equipment disinfection.

Our goal is to prevent healthcare acquired infections, which will ultimately save lives, reduce costs and enhance safety for patients and caregivers. Our clinical data proves best in class efficacy and performance of our platform, and positions PurpleSun to illuminate the healthcare arena.

540922 Mechanisms Contributing to the Formation of "Floating Biofilms" in Staphylococcus Aureus Orthopedic Infections (Invited Talk)

Abstract:

Despite the wide use of antibiotics, breakthrough infections frequently occur after orthopedic surgery. The major pathogen involved with such infections is Staphylococcus aureus. Here, we investigated the mechanisms that contribute to the exceptional recalcitrance of S. aureus to antibiotic treatment during joint infections by an ex-vivo approach using human synovial fluid. We found that even with antibiotic concentrations that far exceed the expected bactericidal levels, S. aureus is not eradicated in synovial fluid (SF).We show that this is due to the fact that in SF, S. aureus forms exceptionally strong biofilm-like aggregates that significantly increase resistance to antibiotic treatment. Screening a transposon bank identified S. aureus fibronectin- and fibrinogen-binding proteins as important for the formation of those macroscopic clumps, suggesting an important role of fibrin-containing clots in the formation of bacterial aggregates during joint infection. While these S. aureus surface proteins are a prerequisite for agglomeration, low activity of the Agr quorum-sensing regulatory system and resulting low production of the phenol-soluble modulin (PSM) surfactant peptides, which function by disrupting interactions of biofilm matrix molecules with the bacterial cell surface, cause agglomerates to grow to exceptional dimensions. Together, our findings support a two-step model of staphylococcal prosthetic joint infection: Interaction of S. aureus surface proteins with host matrix proteins such as fibrin initiates agglomeration; thereafter, the bacterial agglomerates grow to extremely large sizes owing to the specific conditions present in joints. Our findings provide a mechanistic explanation for the reported extreme resistance of joint infection to antibiotic treatment and have important implications for anti-staphylococcal therapeutic strategies.

540853 Bacteria Adhesion Is Mechanosensitive to Polymer Coating Properties (Invited Talk)

Abstract:

Biofilm-related infections are a class of notoriously difficult to treat healthcare-associated infections, which commonly develop on the surface of implanted medical devices. Despite lacking visual, auditory, and olfactory perception, bacteria detect and settle on surfaces, however, how the intrinsic properties of materials affect the initial adhesion microorganisms remains relatively unknown. By unveiling the structure-property relationships between polymer materials and microbial adhesion, we could guide the design of materials *a priori*to resist the adhesion of infection causing microorganisms, such as *Staphylococcus aureus*. In this presentation, I will discuss the effect that fundamental properties of polymer coatings (i.e., molecular architecture, stiffness, and thickness) have on the surface-associated transport of bacteria and on the adhesion of bacteria under quiescent conditions. By decoupling the effects of molecular architecture, stiffness, and thickness from coating chemistry, we have unlocked specific structure-property relationships that can be tailored to control the degree of bacterial adhesion and subsequently, reduce the formation of biofilms.

540784 Invited Talk 3:Â Prospective Technologies Targeting Microbial Biofilm and Its Microenvironment

Abstract:

Biofilm formation is a key factor for a wide range of microorganisms that cause many infections in humans and biofouling in industrial settings. Biofilms are highly structured microbial communities embedded in an extracellular matrix, a multifunctional scaffold that provides adhesion, cohesion and protection to the resident microorganisms. The matrix also provides a diffusion limiting milieu that facilitates the creation of localized chemical and pH microenvironments. These properties contribute to biofilm recalcitrance to mechanical removal and antimicrobial treatment, posing significant challenges for the development of effective antibiofilm therapeutics to control biofilm-associated diseases. This talk will focus on recent approaches developed in our laboratory to disrupt pathogenic biofilms using nano and biotechnology. Using in vitro, ex vivo and in vivo biofilm models with biophysical and super-resolution imaging methods, we have demonstrated how pH-activated nanoparticles or antimicrobials with matrix-degrading activity can target the complex biofilm microenvironment and help prevent biofilm-associated oral diseases. We will also discuss current challenges and future directions for the development of biofilm-targeting therapeutics.

526680 Investigating the Interfacial and Metabolic Properties of Bacteria at Hexadecane-Water Interfaces

Abstract:

Hydrocarbonoclastic bacteria interact with oil-water interfaces in a manner that continues to be unveiled. To degrade hydrocarbons, cells must overcome the challenges of the energies existing at the oil-water interface. The cells become trapped between the two fluid phases that exert interfacial tensions and shear stresses on the cells. We hypothesize that cells develop a protective interfacial layer owing to their ability to metabolize one of the phases. We anticipate that primary oil-eating microbes would exhibit improved intrinsic properties enabling them to withstand the interfacial stress compared to non-indigenous species. Here, we examine the interfacial behavior of *Alcanivorax borkumensis*, a primary oil-eating marine isolate. Using pendant drop elastometry, we characterize the mechanical properties of films of bacteria at the hexadecane-water interfaces (FBI) to understand how cells alter the physicochemical properties of the interfaces. Our results demonstrate that *A. borkumensis* develops a film with viscoelastic properties characterized by frequency dependent storage and loss moduli. We further characterize the interfacial dynamics of *A. borkumensis* under chemical stress to comprehend how cell activity affects the mechanical properties of FBI. Our results demonstrate that cells responding to chemical stress exhibit increased tendencies to alter the interface with time. Furthermore, we correlate data in the pendant drop mode to flat interfaces with particle tracking and microrheology techniques. This two-system approach elucidates the biological implications interfacial bacterial films in a manner relevant to bioremediation.

528984 Complex Liquid EmulsionsÂ and on-Chip Ring ResonatorsÂ for Bacteria Detection

Abstract:

Complex Liquid Emulsions and On-Chip Ring Resonators for Bacteria Detection

Suchol Savagatrup and Timothy M. Swager

Complex liquid emulsions are a class of sensors that translate chemical events into observable morphological responses. Recently, we reported selective and sensitive methods to detect the activity of enzymes and the presence of foodborne pathogens. These methods of transduction rely solely on the changes in the optical transmission of the emulsions upon changes in the morphology, which may limit the generality of the analyte. In this presentation, we report the novel approach to continuously sense and quantify the morphology of these emulsions using on-chip ring resonators. The morphological changes of the droplets are visualized by the characteristic of the transmission spectra of the resonators. Specifically, the shift in the resonant wavelength of the ring resonator is highly sensitive to even the small changes in the localized effective refractive indices induced by the morphological switches. By tuning the compositions of the droplets, we demonstrated the on-chip device capability and sensitivity to detect different classes of analytes. We anticipate the combination of complex liquid emulsions and integrated photonics will provide ultra-compact, cost-effective, and generalized method for other chemical and biological sensing schemes.

535514 Native Airway Mucus Rheology in Health and Patients with Cystic Fibrosis Having Positive or Negative Microbial Culture

Abstract:

In health, airway mucus is composed primarily of long, polymeric mucin proteins as well as some globular proteins and cell debris. Airway dehydration in cystic fibrosis (CF) gives rise to hyperconcentrated, viscoelastic mucus that is difficult for the mucociliary escalator to clear. This unclearable mucus traps pathogens in the lungs leading to inflammation and airway destruction. Because disease onset occurs first in the distal airways, sampling of airway mucus from infant and adolescent patient lungs often relies on the collection of Bronchioalveolar Lavage Fluid (BALF). While the lavage inherently dilutes the airway surface milieu, we show that roughly half (53%) of the mucus-forming mucin biopolymers in CF are present in the airways in insoluble "flakes" or "rafts" that are observable via microscopy. Because such a proportion of airway mucins are insoluble, we hypothesize that the characterization of these flakes will yield important advances in understanding the etiology of CF lung disease. Flakes in BALF present a number of research obstacles mostly arising from the difficulties associated with characterizing the biophysics of heterogeneous samples. To address this, we employ a combination of particle tracking microrheology (PTMR) and Gaussian Mixture Modeling (GMM) in order to isolate and study the biophysical properties and biomedical implications of these flakes.

We collected BALF from 29 patients with CF (age 10.5±4.4 years) and 6 healthy adults. We added 1μm fluorescent polystyrene beads to the samples to probe flake rheology via PTMR. The local complex viscosity (η^*, Pa·s) of BALF samples was computed from mean squared displacement of each individual bead and the generalized Stokes-Einstein equation. Bulk rheology was assessed in the CF samples using a cone and plate rheometer. Samples were scanned for flakes and videos of bead thermal motion in and near the mucus flakes were recorded at 60fps for 30s. Rheological signals were clustered into "watery" and mucus components via GMM.

Comparison between η^* measurements from bulk rheology and the ensemble mean of all beads in each CF patient sample revealed a moderate correlation ($r=0.42$, $p<0.05$). However, comparison of only the mucus component η^* from PTMR enhanced the correlation ($r=0.60$, $p<0.01$), indicating that bulk rheology is more reflective of the thickest components of a heterogeneous sample. Comparing CF samples to healthy adults, weighted η^* (i.e. the sum of the mean η^* in the water and mucus components multiplied by their mixing proportion, p_m) was increased in CF ($\log_{10}(\eta^*)=-0.96$) vs. non-CF (-2.5, $p<0.01$), but the mean of the mucus component was similar. The span of η^* in the mucus component was nearly 3 logs larger in CF samples, and mucin concentration was increased 3,000-fold in CF BALF relative to control samples. Furthermore, the GMM mixing proportion (p_m) of mucus was increased in CF ($p_m=0.83$) vs. non-CF ($p_m=0.27$) samples ($p<0.01$). Interestingly, neither hyperconcentration nor hypersecretion was observed in the subset (n=5) of patients with CF whose BALF was negative for microbial culture. Culture negative samples had weighted $\log_{10}(\eta^*) = -2.8$ and $p_m = 0.23$. The culture positive values of weighted $\log_{10}(\eta^*) = -0.75$ and $p_m = 0.85$ were significantly increased versus the healthy controls ($p<0.01$, both) and the culture negative patients with CF ($p<0.01$, both).

Through a combination of PTMR and Gaussian Mixture Modeling we find that while mucus flakes arise in health and in CF, hyperconcentration only occurs in CF mucus flakes, and orders of magnitude more mucus is found in CF samples. Importantly, increased rheological signal is only found in CF samples that were positive for microbial culture, indicating that infection may be linked to hyperconcentration and hypersecretion.

530457 Engineering Biology to Make Novel Antimicrobials

Abstract:

Proteins perform the cellular tasks required for life. The great variety of their biological activity is due in part to their vast combinatorial space: 20^n, n being the number of amino acids present in any given peptide chain and 20 being the number of natural amino acid monomer building blocks. Yet we do not have the tools to properly engineer these diverse molecules. One approach is to start small: I will present foundational synthetic biology frameworks to rationally develop peptides, tiny proteins that display great sequence diversity but are more amenable than larger molecules to redesign and engineering. My approach is to expand natureâ€™s repertoire to build novel synthetic peptides with extremely useful properties. My overarching vision is to generate a peptide encyclopedia encompassing peptides that target every medically relevant microbe and to devise therapies that nature has not previously discovered. The synthetic peptides that I am developing offer solutions to some of the most pressing unmet clinical challenges we face, including finding strategies for treating antibiotic-resistant infections.

537919 Association with Outer Membrane Vesicles Drastically Alters Bacterial Toxin Activity

Abstract:

Historically, most bacterial protein toxins have been purified and studied in their water-soluble, free form. The mechanisms of these toxins and their roles in disease have been well-established. However, it has recently been observed that many bacterial toxins can also be released in association with outer membrane vesicles (OMVs), which are produced from the outer membrane of Gram negative bacteria. Our lab has studied the activity of two bacterial toxins that are associated with OMVs and found significant differences in the delivery and activity of both when they are associated with OMVs compared to when they are free. *Vibrio cholerae*, the causative agent of cholera, secretes cholera toxin (CT), which binds to the ganglioside GM1 on host cells to initiate its toxic mechanism. We have discovered that most of the secreted CT is actually associated with OMVs, not in its free form, and in its OMV-associated form, the toxin does not bind to GM1. This finding highlights the need to understand delivery of this toxin in its OMV-associated form, as inhibitors of GM1 binding by CT are unlikely to be effective anti-virulence strategies. Similarly, the oral bacterium, *Aggregatibacter actinomycetemcomitans*, secretes a leukotoxin (LtxA) in a water-soluble form. LtxA recognizes cholesterol and an integrin receptor, LFA-1, on host cells. Because LFA-1 is only expressed by human immune cells, this toxin (in its water-soluble form) has been reported to be active only against these specific cells. However, OMV-associated LtxA is delivered to cells in an LFA-1-independent manner and is active in cells that do not express LFA-1. Importantly, this finding demonstrates that the activity of this toxin is much greater than has been suggested in the past. Together, our results indicate that bacteria package protein toxins in vesicles to enable long-distance delivery and to expand the range of activity of the toxin. In order to fully characterize the virulence of pathogenic organisms, it is essential to study these proteins in their vesicle-associated form, as well as their free form.

541647 Can Soft-Matter Mechanics Provide New Avenues for Remediating (and even preventing!) Biofilm Infections?

Abstract:

Biofilms are systems of microbes that are bound to each other, and often to an external material, by a matrix of biopolymers and proteins. Therefore, biofilms are characterized by mechanical properties that are not present for systems of equivalent microbes in the non-biofilm state, with individual organisms suspended in fluid. We investigate how these mechanical properties contribute to the pathogenesis of biofilms, and how mechanical properties of the biofilm and/or its environment might be altered to make biofilms easier to clear, and even to prevent biofilms from developing.

We recently found that chronic, years-long biofilm infections evolve to change the production of matrix materials in such a way as to increase or maintain the mechanical robustness of the biofilm; we use shear rheology to measure mechanical robustness in terms of elasticity, yield strain and stress, and the energy input required to cause the biofilm to yield. More recently, we have begun investigations of how these mechanical properties of biofilms could impact the ability of the immune system to clear biofilms. Our preliminary results show that varying the elasticity of abiotic gels over the range measured for clinical biofilms has greatly influences how likely it is that neutrophils (phagocytic immune cells) will break off and engulf parts of the gel. This suggests that weakening relevant mechanical characteristics of biofilms may help to promote their clearance by neutrophils, which are the body's first responders and the primary defenders against neutrophils. We will present early results on approaches to disrupt biofilm mechanics by attacking the interactions between specific matrix materials.

541736 Mechanisms of Bacterial Biofilm Growth and Biofilm-Virus Interactions

Abstract:

On most moist surfaces, bacteria grow into communities termed biofilms, which display strongly increased resistance to antibiotics, and constitute the most abundant form of biomass on Earth. I will first present new microscopy and imaging techniques we recently developed to follow dynamical processes in biofilms at the single-cell level. I will then proceed to show how we applied this novel imaging technology in combination with genetics and molecular techniques to discover mechanisms of basic processes in biofilm development: I will show how individual cells within biofilms interact with each other, how cells disperse from biofilms, and how bacterial biofilms interact with bacteriophage viruses, which are the most frequently encountered predators of bacteria in the environment. This work on biofilm dynamics at the single cell level promises many new discoveries for preventing biofilm growth on environmental and biomedical interfaces.

541645 The Biophysics of Bacterial Biofilms Facilitate Surface Survival in Moving Fluids but May Reveal an Achilles Heel

Abstract:

Bacterial biofilms are microscopic assemblages of bacterial cells usually attached to a surface and held together by a self-produced extracellular polymeric slime (EPS) matrix. Biofilms are ubiquitous in the natural environment and are highly problematic in industry and medicine where they cause corrosion, fouling, contamination and chronic medical and dental infections. The EPS matrix is chemically complex and is composed of polysaccharides, lipids, proteins and DNA and protects the bacteria within from antibiotics chemical challenges and host immunity. Mass transfer through the EPS is dominated by diffusion which, allows the development of gradients in nutrients, waste products and cell signals. The bulk mechanical properties of biofilms play an important role in survival by allowing the biofilm to respond to fluid shear stresses over very short (ms) and very long (days to weeks) time scales. Creep and relaxation tests show that biofilms behave as viscoelastic liquids however and recent observations of high velocity impacts with water droplets suggest they rapidly form interfacial instabilities allowing them to flow over surfaces with velocities of meters per second. Generating turbulence in the biofilm might explain enhanced killing by antimicrobial agents in dental biofilms and have broader applications in biofilm control. A better understanding of how bacterial biofilms respond to fluid flow provides new opportunities to develop more effective control strategies.

541646 Biofunctionalization of Implants through Thin Films

Abstract:

Control of the tissue-implant interface is integral to the success of the implant, and ultimately the health of the patient. Through a series of projects on multiple types of implant substrates, the Gawalt Lab has developed a system for immobilizing multiple bioactive molecules on the surface to control the interface using strongly adhered linkers and basic organic chemistry. For example, ceramic surfaces have been modified as bone fillers to enhance bone growth and reduce infection with either one or two molecules, while stent materials are being modified to reduce platelet aggregation and stimulate endothelization. Recently, a triple-functionalized surface has been completed on orthopedic implant materials to simultaneously enhance bone growth, while reducing bacterial adhesion. This new system is the first system with three bioactive molecules in a coating.

541773 Infection-Resisting Biomaterials

Abstract:

The synthetic surfaces associated with tissue-contacting biomedical devices are attractive to many bacteria, because these surfaces provide anchoring sites, often well protected from the innate immune response, in a moist environment with abundant access to nutrients. Clinically, bacterial colonization of device surfaces is significant, because a series of revision surgeries is often required to resolve the resulting implant-associated infection. The impact on both the patient and the health-care system can be substantial. While better surgical practice can mitigate the chances of infection, eliminating bacterial surface colonization is in large measure an engineering problem. One solution has been to incorporate antimicrobials into the device. These then elute into the surrounding physiological system over extended time periods after surgery. This approach, while FDA-approved in a small number of applications such as antibiotic-loaded bone cement, delivers antibiotic whether it is needed or not, and it thus runs counter to the rapidly increasing concerns about the overuse of antibiotics and the ongoing evolution of antibiotic-resistant bacterial strains. An alternate concept is that of self-defensive surfaces. These sequester antimicrobials until their release in response to a bacterial trigger. Sequestered antimicrobials are present only in very small amounts, and they are released only if needed. Consequently, they do not contribute to the development of resistance. This presentation describes work using anionic microgels to create self-defensive surfaces. These charged microgels can be electrostatically deposited onto surfaces and subsequently loaded with small-molecule cationic antimicrobials by complexation processes. We have identified conditions where complexed antimicrobials remain sequestered for several weeks despite the fact that a diffusion-based elution process would take only seconds. These sequestered antimicrobials can, however, be released when closely approached by bacteria. Contact creates a steep chemical-potential gradient that drives antimicrobial transfer from the microgels to the bacterial envelope where there is a high concentration of negative charge. This transfer kills the bacteria. Metabolizing tissue cells in contact with the same microgel-modified surfaces are, however, unable to trigger antimicrobial release, indicating that the contact-transfer mechanism is specific to bacteria. Such specificity is critical for designing future biomaterials that not only promote healing but simultaneously resist infection.

AUTHOR INDEX